New England Adamant Co.

New England Adamant Co.

Manufacturers of Adamant Wall Plaster

New England Adamant Co.

New England Adamant Co.
Manufacturers of Adamant Wall Plaster

ISBN/EAN: 9783337022020

Printed in Europe, USA, Canada, Australia, Japan

Cover: Foto ©berggeist007 / pixelio.de

More available books at **www.hansebooks.com**

"ADAMANT,"

New England Adamant Co.

MANUFACTURERS OF

Adamant Wall Plaster.

WORKS :

FIRST ST., FOOT OF E, SOUTH BOSTON.

OFFICE :

21 FEDERAL ST., BOSTON, MASS.

Opinions of the Architectural Press.

From the Northwestern Architect, Minneapolis, Minn.

"As is compatible with this progressive period, even wall plaster has succumbed to inventive genius, and 'the day is nearly done' when our handsomely decorated walls and ceilings shall be ruined by reason of the utter unreliability of the old style mortar. To meet this demand for a non-porous plaster—Adamant has appeared on the market—and is now being manufactured in nearly every State in the Union."

From the Scientific American, August, 1887.

"Some of the advantages Adamant possesses above the other plaster is the extreme hardness of the wall. It will neither crack or crumble. *It is a strong support to a building.* It does not swell timbers by an absorption of moisture, *as the material itself absorbs most of it.* It dries so quickly after it is applied to the wall, that in a few hours *frost will have no effect on it.* We are convinced that this Adamant is destined to revolutionize the business of house plastering."

American Architect.

In this age of improvement it seems strange that men should so long have been confined to the use of lime, sand and hair for making interior walls and ceilings, especially since walls so made have proved very weak and unsatisfactory, and many know from sad experience what destruction, annoyance and loss ensue, when, from some slight cause, its own weight or rottenness, down come ceilings about their ears with the inevitable result of damage, dust and confusion. To all who have had such experiences, the advent of Adamant Wall Plaster will be a matter of great interest. It combines all the qualities of a perfect plaster, viz.: hardness, ease of application, density, resistance to fire, durability, non-swelling and shrinking.

From the Builder and Decorator, Philadelphia, Pa.

"Nothing in house building has so long withstood the 'march of improvement,' notwithstanding it has failed to give a substantial wall, while many houses become defaced and old in a few months on account of broken walls and ceilings." Adamant obviates all this, and is the coming wall plaster.

NEW ENGLAND ADAMANT CO.,

MANUFACTURERS OF

Adamant ÷ Wall ÷ Plaster.

CAPITAL, $150,000.

N. J. BROCKWAY, General Manager.

BOSTON, MASS.

ADAMANT.

A wall has no business to be "dead." It ought to have members in its make, and purpose in its existence, like an organized creature, and to answer its ends in a living and energetic way; and it is only when we do not choose to put any strength nor organization into it that it offends us by its deadness.—*John Ruskin.*

ADAMANT WALL PLASTER.

Adamant Wall Plaster, as its name implies, is a material designed to produce a hard and practically indestructible interior wall and ceiling.

Within five years Adamant has made for itself a wide reputation, entirely on its merits, having been applied in tens of thousands of buildings, both public and private, in Massachusetts and other States.

The demand for Adamant is rapidly increasing. It now has an established place as a standard building material.

The plastering has long been the poorest and most unsatisfactory material used in the construction of buildings, and a good interior wall is the exception where lime plaster is used. A slight concussion breaks the surface of the wall, and patches, holes and defaced decorations offend the eye. Where lime plaster is used it is not an uncommon occurrence to have the whole or a part of the ceiling fall. To avoid this, many churches, school buildings,

3

stores, etc , have been ceiled with wood, thereby largely increasing the cost.

By using Adamant the necessity for wooden ceilings is entirely obviated, as it adheres stron ly to any substance, and can not be removed without considerable effort.

In recent years numberless improvements have been made in building materials generally, but until the introduction of Adamant, a superior wall at moderate cost was unobtainable.

In the ordinary way of making and applying common plaster, results are very uncertain. The varying qualities of lime and sand render fixed proportions unsafe. The plaster must be mixed weeks before it is used, so that the lime may become thoroughly slaked. Frequently the hair is destroyed by the action of the lime, and a wall made of such material will be " rotten." Much care must be observed about the drying of one coat before the next is applied. The large quantity of water used in common plastering, to the injury of the wood-work, the time required to dry the building, and in winter the expense of fuel for that purpose, are among the disadvantages inherent in its use.

All these difficulties are avoided where Adamant Wall Plaster is used. It is a chemical composition, and in a few hours after being applied becomes very hard and capable of resisting, intact, all the ordinary casualities that prove so destructive to common plaster. It is a dry material, shipped in bags, ready for use by simply mixing with water, and is applied in the usual manner. If kept dry it will not deteriorate with age.

Among its many advantages Adamant saves time and labor both in preparing and in applying; it is cleanly and easy to work; it avoids saturating the timbers with water and the consequent swelling and shrinking; it saves waiting weeks for rooms to dry out, and they can be safely occupied immediately after finishing.

4

Carpenters need not move out while the plastering goes on, but can continue work on the same floor with the plasterers. Of course, until the Adamant has set, walls should not be jarred.

Its resonant properties make it particularly valuable in churches, halls, opera houses, and all public places.

In fire-resisting qualities it is superior to any other plaster made, and this consideration alone should have great weight with those who are building or preparing to build.

Adamant does not crack or shrink ; rats do not gnaw through it, nor will it harbor vermin, noxious gases, or germs of disease, like common plaster, because it is smooth, dense and hard, instead of porous. It places much less weight on the building than lime plaster, and instead of being a dead weight it contributes strength.

It has great adhesiveness and considerable elasticity, therefore for ceilings which are liable to " spring " it is invaluable. It can be frescoed, papered, painted or treated with any desired finish within four or five days after the application of the last coat.

No one who intends building substantially and economically can afford to use common plaster,.when, for a moderate additional first cost, he can put on Adamant and obtain a solid wall that will not be injured every time it is touched by furniture ; that is cooler in summer and warmer in winter, and in every respect superior.

REMARKS.

Adamant is sold in bags holding half barrel each. We expect them returned in good order as soon as emptied. Bags lost, destroyed, or not returned, *must be paid for.* They should be kept dry, and can be packed in strong paper, or bagged (using some of the same for that purpose), and returned by freight. Each bundle must be tagged, and the sender's name and number of bags returned plainly marked on tag. Shipping receipt must be sent in all cases.

We make the following kinds of material:

Adamant No. 2, for base or first coat.
" No. 1X, or float finish.
" No. 1XXX, or white trowel finish.

If walls and ceilings are lathed as directed, 5 to 6 barrels of No. 2, and 2 barrels of finishing material will cover 100 square yards. Where brick walls are reasonably straight it will take on an average 8 to 10 barrels of No. 2 for 100 square yards.

DIRECTIONS.

In preparing a building for Adamant Wall Plaster the grounds should be put on ⅝ of an inch, which will give a full quarter inch of material above lath, and the laths should be not less than ⅛ inch apart on side walls, and from ⅛ to ¼ of an inch on ceilings. If further apart there will be loss of material, and if nearer together room is not allowed for the swelling of lath, and in such cases there is danger of buckling or bulging. See that lath nails are well driven in, heads flush with lath. Brick, tile, fireproofing and all other porous material should be thoroughly wet before applying Adamant. If the suction is so great as to draw the water out of the material before it has set, the strength will be greatly reduced.

MIXING.

Use a common hoe and a water-tight mixing box. Mix thin with water at first, in order to dissolve the chemical ingredients and get the full strength, then add sufficient dry material to bring it to the right consistency to work smoothly and well. Do not mix more than can be used in one and a half hours. Keep box and tools thoroughly clean.

Put on No. 2 for first coat, spreading freely, but little pressure being needed; then straighten with feather or straight-edge or darby. When the latter is used the wall should be first slashed with water or the darby wet. With good summer weather or with heat in winter this wall will be ready for second coat of 1XXX in about three days. It can be applied sooner, but more satisfactory results will be obtained by waiting that length of time.

It is a good plan, especially in the winter and cooler months, to put on first coat for one entire floor, when trowel finish is to be applied, then follow in the same order with finish. This avoids staining from sap lath.

To apply No. 1XXX the under coat must be dry. First time over apply very thin, grinding it in just filling the pores. Allow it to draw a few minutes to avoid blistering, for as the moisture is drawn into the dry coat the air is expelled. The thin coat also serves to stop suction sufficiently, and allows plasterer to lay it on next time over perfectly level. The last time over (for we go over it three times), thin up the Adamant on the board, just so it can be handled on the hawk, filling in all cat faces and other imperfections, and going all over with this thin material, finishing nearly as possible. After it has drawn awhile trowel off, using damp brush; (dip brush and sling water out of it before using). No more labor should be put on after good surface is obtained, as the material does not chip crack or fire crack. Finally brush with dry brush. In finishing Adamant use little water as possible, and make few joinings by working top and bottom together.

If a float finish is desired (which is the cheapest and most

7

rapidly finished wall, and well adapted for painting), apply No. 1X, and float, using little water as possible. The float finish should be applied within 24 hours after the application of the first coat.

We strongly recommend stippled work in preference to float finish, the work is easier and result more satisfactory. In this class of work we use No. 2, for first coat on lath, left smooth as possible and, while comparatively green, putting on thin coat of No. 1X, which is stippled at once either fine or coarse as desired, using ordinary stippling brush.

Frost will not harm Adamant after it has been on ten hours. The second coat, however, should never be applied while there is frost in the first coat.

Alterations and repairs can·be done nicely and expeditiously with Adamant. Wet the old lath, also the edges of old mortar, to reduce suction.

Damp walls are effectually remedied by applying Adamant and then painting or oiiing.

All walls should be glue sized before being papered.

PRICES.

Delivered on cars at Boston, Mass., in returnable bags.

Adamant No. 2, for base or first coat on lath, per bbl. (2 bags), 260 lbs., ... $2.00
A lamant No. 1X, or float finish, per bbl. (2 bags), 280 lbs., ... 3.00
Adamant No. 1XXX, or white trowel finish, per bbl. (2 bags), 220 lbs., ... ·. 3.50

When ordering Adamant, be sure and state the kind or finish required.

SUMMARY.

ADAMANT is a cement or an artificial stone, is very hard and will last as long as the building.

ADAMANT places much less weight on a building than lime mortar; and instead of being a dead weight, contributes great strength.

ADAMANT does away with the warping and shrinking of doors and casings, and the building is not saturated with water as it necessarily must be when common plaster is used.

ADAMANT adheres firmly to laths, brick, stone or iron, has great elasticity, and for ceilings liable to spring is invaluable.

ADAMANT costs originally but little more than lime plaster, and in the end its superior qualities make it immeasurably cheaper.

ADAMANT is especially in demand on account of the great saving of time by its use. Wood finish can be safely applied in one week from time plastering is commenced and a house occupied as soon as completed without danger from moisture.

ADAMANT does not crack or fall off, even in the case of leakages.

ADAMANT will not crumble.

ADAMANT in the hands of a good plasterer is easily applied and will do credit alike to itself and the workmen.

ADAMANT is the only material with which repairing can be done neatly and "to stay."

ADAMANT is the best material with which to make a fire-proof wall, by applying on wire lath, brick or terra-cotta.

ADAMANT will hold fire in check when used behind basings and other wood-work.

ADAMANT is strongly endorsed by painters. First, because it does not change their colors. Second, as it will not crack, they can apply the richest decorations with perfect confidence.

ADAMANT is appreciated by paper-hangers, since it enables them to do first-class work.

ADAMANT is not porous, as is common plaster, and therefore does not absorb gases or become the nucleus of germs of disease; hence its sanitary qualities have commended it for use in many large hospitals.

ADAMANT can be cut clean to a straight edge. Registers or stove-pipe thimbles can be put in without removing extra plaster.

ADAMANT, because of its density and the fact of its being a non-conductor of heat, makes a building warmer in winter and cooler in summer.

ADAMANT is now being manufactured by thirty companies, operating in every State of the Union, and has been applied on tens of thousands of buildings both public and private, with universal satisfaction to architects, owners and builders.

ADAMANT needs only to be used to be appreciated.

SUGGESTIONS

For Specifications for Adamant Plastering.

Lath Work.—Grounds to be put on $\frac{5}{8}$ of an inch, so that there shall be not less than $\frac{1}{4}$ inch of Adamant above the lath. Laths to be of best pine or spruce (when white finish is desired), free from sap and bark, and to be put on from $\frac{1}{8}$ to $\frac{3}{16}$ of an inch apart.

Brick Walls, Tile or Fire-Proofing shall first be thoroughly wet, to reduce suction, and then to receive a coat of No. 2, which shall be straightened and brought to an even surface. When in proper condition to be finished as follows:

When a White, Smooth Surface is desired. — Ceilings and walls to receive a finishing coat of Adamant No. 1XXX, troweled to a true and smooth surface.

For a Float or Sand Finish. — All walls to receive a finishing coat of Adamant No. 1X, floated to a true and even surface, free from float marks and cat faces, the floating to be done with the use of as little water as possible, to avoid killing the surface.

For Stippled Work. — Walls to receive a finish of Adamant No. 1X, to be applied evenly when under coat is in proper condition, and stippled rather coarsely with ordinary stippling brush.

DETAILED INFORMATION.

We manufacture three kinds of Adamant as stated on page 6. Our price list on page 8 is subject to a discount of 50 per cent. when bought in quantities. In preparing a building for Adamant, follow instructions given on page 6. It can, however, be applied on a building prepared for lime mortar, thicker grounds and open lathing, but it takes more stock.

We furnish Adamant the same as the Lime Dealer does the lime, sand and hair. We cannot quote or estimate cost of wall complete as we do not know the price of labor in the various cities and towns which is an important factor when the entire cost of wall is taken into consideration. For a building prepared as stated the following is a liberal estimate for one hundred square yards, surface on wood lathing :

From five to six barrels of No. 2, first coat, and two barrels, second coat (No. 1X if float finish is wanted for a paint or fresco surface, or No. 1XXX if fine trowel finish is desired).

Adamant commends itself to the public, in the following respects: First, its superiority over lime. sand and hair is unquestionable, see pages 9 and 10. Second, the great saving of time. It is actually cheaper, even so far as dollars and cents are concerned, to say nothing abont real value, when used in the construction of business blocks and tenement houses.

A building can be delivered and occupied from four to six weeks quicker than when lime mortar is used and the rent for this time saved will much more than cover the extra cost. Hence an Adamant wall with all its superior qualities is secured not only at the same cost of a lime wall but actually less.

The claims of Adamant have been conceded bv those who have used it, among whom are the leading architects in the principal cities of our United States, contracting builders by the score and owners innumerable.

We trust that the merits of Adamant, as stated in the preceding pages, will be carefully considered and thoroughly investigated, assuring the public that these claims are unquestionably substantiated in the following list of testimonials. If, after carefully perusing these pages, all questions are not clearly answered or if there are any other points on which more detailed information is desired, kindly embody the same in a letter to us.

NEW ENGLAND ADAMANT CO.

IMPORTANT.

TACOMA, WASHINGTON, Jan. 10, 1890.

W. E. Sharps, Manager, The Adamant Plaster Mfg. Co., Tacoma, Wash.

DEAR SIR :— Referring to the fire of January 4th, 1890, at the Fannie C. Paddock Hospital, Tacoma, we take pleasure in saying that in **our opinion** the restriction of the fire to the building in which it occurred was largely owing to the **Adamant Plaster** on the inside walls of said building. **The fire-resisting qualities** of your plaster were very clearly shown by **this severe test ;** and representing, as we did, **a large** amount of insurance on the hospital building, we desire, in behalf of the companies interested, to express **our appreciation of the merits** of your invention. E. NILES,

Special Agent and Adjuster, Union Insurance Co. of California.

DELPRAT & BALL, Resident Agents.

An Extraordinary Fire-Test.

Extract from Birmingham, England, Daily Times, 6th April, 1889.)

A FIRE RESISTING PLASTER —With the object of demonstrating the fire resisting properties of Adamant Plaster, a manufacture of the Adamant Company, Limited, Birmingham, some interesting experiments were given on Thursday on the unoccupied space in Corporation street, opposite the New Law Courts. Superintendent Tozer of the Birmingham Fire Brigade conducted the experiments, and Alderman White and a few of the public were present at the invitation of the company. Two square structures representing rooms, with chimney place and entrance, had been erected, the interior of one being lined with the Patent Adamant Plaster, whilst the second was constructed of the ordinary hair mortar. The ceiling of the first structure was composed of Adamant pure and simple, and that of the second of the ordinary plaster and laths. Huge piles of small timber and shavings were placed in the rooms, and at a given signal the heaps were ignited simultaneously. In about fifteen minutes the second structure showed signs of giving, the laths at the sides first igniting, whilst the plaster commenced to drop from the walls and ceiling. Five minutes later the ceiling caught, and the greater portion fell in with a crash, the fire continuing to burn fiercely. THE BUILDING OF ADAMANT WITHSTOOD THE FURIOUS HEAT. So far, the value of the Adamant was proved beyond doubt.

(Official Report of the Superintendent of the Birmingham Fire Brigade.)

Chief Fire Station, BIRMINGHAM, 10th April, 1889.

The Superintendent of the Fire Brigade begs to report that on the 3d inst. he witnessed a trial made by the Adamant Company, of Commercial street, Birmingham, under his supervision, to demonstrate the fire-resisting properties of Adamant as compared with ordinary plaster. The trial was conducted publicly on the land in Corporation Street, opposite the Assize Courts, and was a severe one, the result proving —

1. That as a fire-resisting material, Adamant is far superior to ordinary plaster.

2. That the patent system of using the Adamant for ceilings and partitions is infinitely superior to the ordinary method of lath and plaster, and the pugging usually adopted.

A. R. TOZER, Superintendent.

13

Report of the Committee appointed by the Kansas City Academy of Science to investigate Adamant.

Kansas City Academy of Science.

GENTLEMEN:—Your committee appointed to investigate the properties of Adamant Wall Plaster would report as follows:

On Friday, August 23d, the committee visited the factory of the Adamant Plaster Company. After viewing the preparation of the material, a number of tests were made upon surfaces of plaster which had been laid on lath and brick. The following facts were established:

A strong blow delivered with a hammer or stick of hard wood crushed the plaster at the point of concussion without producing radiating cracks nor chipping off the surrounding plaster.

A nail could be driven through the plaster into the lateral centre of a lath without breaking the clinches of the plaster.

A nail so driven could be withdrawn with a minimum breakage of the surrounding plaster.

The finishing coat shown us presented a very smooth and polished surface.

The brown coat shown us which had been put on twenty-four hours previous, was firm and well set, and offered a strong resistance to blows.

In all the qualities mentioned it far excelled ordinary plaster.

In all respects it is superior to any plaster which has been brought to the notice of any member of the committee.

Dr. R. Wood Brown and Mr. Charles W. Dawson were appointed a sub-committee to further investigate Adamant.

Dr. R. WOOD BROWN, Chairman.

FREDERICK McINTOSH, ⎫
E. T. KLEIN, ⎬ Committee.
CHARLES W. DAWSON, ⎭

The sub-committee reports as follows:—

We have established the following facts:

The labor of working Adamant Plaster is about the same as for ordinary plaster.

Adamant Plaster sets hard and firm in from six to twelve hours.

The tensile strength of the rough coat (such as is used for laths) when it has set for twenty-four hours, is One Hundred and Fifteen Pounds to Square Inch. We consider tensile strength entirely satisfactory.

Dr. R. WOOD BROWN, Chairman.
CHARLES W. DAWSON.

Prominent Fire Insurance Men's Opinion.

BOSTON MANUFACTURERS MUTUAL FIRE INSURANCE CO.
31 MILK STREET, BOSTON, July 8th, 1890.

MESSRS. BURLINGTON WOOLEN CO.

DEAR SIRS: From what we already know of the material known as Adamant Plaster, which is offered you by the New England Adamant Company, we should not hesitate to recommend you to use it wherever it can be put on ceilings, on wire lath, or made use of in any other way to retard the effect of, or the spread of fire.

We have as yet no positive experience under our own observation or in our own risks of its service, but it possesses the merits which are required in dangerous places. It is strong; it is non-heat conducting; it can be readily applied without any serious stoppage to machinery; it sets very quickly, and so far as we know, is not readily broken or shaken off by the vibration of machinery. I am now making a thorough investigation of the subject with a view to offering the makers the opportunity to send their circular under our wrapper to our members, so that they may all be informed upon the subject.

Yours very truly,   EDW. ATKINSON, Prest.

FALL RIVER MANUFACTURERS MUTUAL INSURANCE CO.
FALL RIVER, MASS., July 3d, 1890.

New England Adamant Co.:

DEAR SIRS: The experience of the Manufacturers Mutual Insurance Co.'s has demonstrated the importance of having no wooden ceilings or partitions exposed in departments most subject to fires, such as Picker rooms, the Dust rooms of Cotton Mills, and kindred departments in other classes of manufacturing establishments.

Your "Adamant Wall Plaster," laid on wire lathing, has recently been applied in a considerable number of picker rooms and dust rooms, as well as stairway enclosures with great facility and excellent results.

It hardens within a few hours so that if applied in the afternoon or first half of the night the usual operations of the departments can be pursued on the following morning with no injury to the plastering or hinderance to the work. It is much more tough and durable under blows or battering than ordinary Lime Mortar. We most heartily commend it for such purposes.   Yours truly,

THOS. J. BORDEN, Prest.

15

Testimonials.

No better evidence can here be given of the great superiority of Adamant than the testimony of those who have used it. We therefore ask you to examine the following letters from some of our many patrons, to satisfy yourself of the validity of our claims:

SPRINGFIELD, MASS., May 28, 1890.

N. E. Adamant Co., Boston, Mass.

GENTLEMEN:—I have used Adamant Wall Plaster on several buildings and indorse it **without qualification.** I am heartily glad that at last architects have a material to use which will give them a wall such as was absolutely impossible to secure with any plastering material prior to the advent of Adamant Wall Plaster. **We have stood the nonsense of rotten lime walls, falling ceilings, plastering full of cracks, crumbling and running out at the slightest touch, waiting weeks of time to dry out, until patience has ceased to be a virtue.** Your material has been carefully tested and has proven to be just what you claim. Very hard, but very elastic. I have yet to see a single instance where the carpenters have broken the surface in putting up their wood finish. **Its adhesive qualities are wonderful;** it being impossible to remove it from the laths, or in fact from anything to which it may be applied, without digging it off inch by inch. **Its remarkable fireproof qualities** will certainly **reduce the rates of insurance.** Soaking the ceilings or walls does not cleave it off or impair its strength one particle. I have the courage of my convictions, and my experience has proven that I am warranted in making Adamant a permanent feature in my specifications.   Yours truly,

J. M. CURRIER, Architect.

BOSTON, MASS., August 23, 1889.

New England Adamant Co., Boston, Mass.

GENTLEMEN:—During the past year I have used your Adamant Plaster in the Niles Building, School street, and can with pleasure endorse it. **Not only does its use save about three weeks of valuable time, but it gives a superior surface for decoration.**

Yours respectfully,   WILLARD M. BACON, Architect.

BOSTON, MASS., April 27, 1889.

New England Adamant Co., Boston, Mass.

GENTLEMEN:—I have been using "Adamant Walls" quite extensively in my work during the past year, and with perfect satisfaction to my clients and myself. Their **great strength and elasticity, the weeks of time saved** by quick hardening and drying, and **their fire-proof qualities**, are among their strong points. Wood finish can be safely applied within **a week**, and they can be papered or painted within a few days after completion. I can cordially endorse the Adamant Walls, and shall use them in my work whenever possible.

Yours very truly.   W. R. EMERSON, Architect.

Fuller & Delano, Architects, 452 Main St.

WORCESTER, May 24, 1890.

New England Adamant Co., Boston, Mass.

GENTLEMEN:—We have specified and used the Adamant plastering in several instances with good success. The last work finished with Adamant was P. W. Moen's house on Elm street, Worcester, where we used both a troweled and rough floated surface on wire lathing. **We have no hesitation in saying it is the best job of plastering we have ever had in an experience of twelve years.** The surface is hard and uniform, of even color, and much stronger than lime plastering, and saved us about three weeks in time.   Yours truly,

FULLER & DELANO, Architects.

BOSTON, MASS., January 21, 1888.

Bay State Adamant Co., N. J. Brockway, Manager.

DEAR SIR:—The Adamant Wall Plastering you put on for me in Wellesley is a great success. Although the house had been prepared for ordinary lime and hair mortar, with very coarse lathing and three-fourths grounds, and was in a cold corner of the building where the temperature was not many degrees above the freezing point, **your first coat was hard in forty-eight hours and dry enough for the finish coat,** while the ordinary lime mortar in an adjoining room that had been on ten days was still soft. **The walls are extremely hard,** and it seemed impossible to crack the finish coat by nailing the casings. The plumber, having had occasion to cut some holes, said to me that he wished he could always work on such walls. I should have no hesitation in specifying it.   Respectfully,

W. FRANK HURD, Architect, 35 Congress St.

17

MANCHESTER, N. H., July 29, 1890.

New England Adamant Co., Boston, Mass.

GENTLEMEN :—R plying to your favor requesting my opinion of the Adamant Wall Plaster used throughout Mr. Geo. A. Rollins' fine residence at Nashua, N. H., would say that **it is rightly named "Adamant" or a "Perfect Wall Plaster."** In all my experience as an architect **I have never seen a wall which would compare with yours,** either in hardness, strength, beauty of finish, or absence of cracks. **Your wall is as marble** while a lime mortar wall, at its best, is rotten and a dead weight to a building.

One of the ceilings in this house was flooded from a leak in the roof, and the plastering was not injured in the least. Had it been lime mortar it would have fallen or cracked so that it would have been necessary to remove it. The slight additional first cost of Adamant should not argue against it when contracts are being made for private residences, business blocks, public buildings, or tenement houses, as **it is of the utmost importance that the plastering should be indestructible.**

My convictions, hence my advice, especially to my clients, is, if their estimates exceed the amount intended to be put into a building and it becomes necessary to cut down, **Adamant should never be sacrificed.** Your walls are your best testimonials, and Mr. Shattuck of Nashua **after carefully examining** Mr. Rollins' plastering has contracted to have Adamant used throughout his new residence **in place of lime mortar** as originally specified.   Yours very truly,

WM. M. BUTTERFIELD, Architect.

RAND & TAYLOR, ARCHITECTS,

28 SCHOOL STREET, BOSTON, July 26, 1890.

Mr. N. J. Brockway.

DEAR SIR :—Regarding the plastering for the **new Mary Hitchcock Hospital at Hanover, N. H.,** we would say that we have taken the trouble to investigate the various claims for and against your Adamant, and **have presented the same to the donor of this building and the Medical Experts,** and as a result we have let the specifications stand as originally drawn up, **calling for Adamant throughout,** and we shall look to you to see that the proper material is used and in a proper way.   Respectfully yours,

RAND & TAYLOR.

Boston, Mass., January 24, 1889.

Bay State Adamant Co., Boston, Mass.

Gentlemen:—In reply to your inquiry as to my opinion of "Adamant Wall Plaster," I would say that **after a year's trial,** having subjected it to the most severe tests, and often under very unfavorable conditions, **I give it my unqualified endorsement in every particular.** Its strength is remarkable, and **its quick hardening and drying qualities** recommend it strongly to builders. It can be applied easily by any good plasterer, and my men **can spread as much in a day as of lime mortar.** In a word, it is all you recommend it to be, and it cannot fail, in my judgment, to become a great success, and one of the standard building materials. Yours truly, DAVID McINTOSH,

Contracting Plasterer, 164 Devonshire St.

Boston, Mass., September 3, 1889.

New England Adamant Co., 21 Federal Street.

Gentlemen:—Regarding the "Adamant" Plaster which was put on my house, I am very happy to say that it is perfectly satisfactory in every respect. I certainly expect a warm and comfortable house in winter. **The carpenters were all loud in its praise,** and they lost little or no time waiting for it to dry out. The ceiling of my dry-house for drying lumber is also lasting very well, and is far superior to the old plaster which it replaced. The ceiling is, as you well know, constantly subjected to hard knocks while the lumber is going in and out of the dry-house.

The job of plastering on my house has been pronounced by judges **"one of the finest jobs of plastering they ever saw,"** and certainly is a first-class one. If my recommendation of so good a thing is of any value, you are very welcome to it.

Very truly yours, ELIOT B. MAYO.

Boston, Mass., August 30, 1889.

New England Adamant Co., 21 Federal Street.

Gentlemen:—I occupied my new building, 516 Washington street, **one month earlier** by using "Adamant" than I could have done had I used common lime mortar for plastering. Floors were laid, window and door casings were put on in fourth story while plastering was being done in the first. This **saving of time alone should cause "Adamant" to be universally used in the construction of business blocks.** The walls and ceilings are very hard. No patching to be done after carpenters left, as they did not even mar the surface in nailing on the wood finish. I cheerfully attest to its superior qualities.

Yours respectfully, R. H. WHITE.

BOSTON, MASS., September 7, 1888.

Bay State Adamant Co.

GENTLEMEN —The Adamant Plaster used upon the walls of our banking-room has proved to be all that your Mr. Brockway recommended it to be. For walls **liable to hard usage,** we should recommend it as being **superior to any other plaster.**

Yours respectfully, THOMAS N. HART,

President Mount Vernon National Bank.

BOSTON, MASS., April 10, 1889.

New England Adamant Co.. Boston, Mass.

GENTLEMEN:—In reply to your inquiry would say, we used Adamant walls on our new Café at Sorento, Me., in June of 1888. **They were put on complete in seven days, and wood-work followed as fast as flat was completed.** There has been no trouble with swelling or shrinkage on account of moisture. We saved **weeks of time** by using Adamant, and the Café was ready for opening in about two weeks from the time we commenced plastering. The walls are all and more than you clam for them, **are very hard, do not crack, and have withstood the cold and severe storms of the past winter without a blemish.** We can recommend them as being in every way superior to any lime wall we have ever seen. In closing we can say that we heartily endorse your walls, and know from our own experience that they have no equal. Wishing you the success which your enterprise deserves, we are, Yours very truly,

FRENCHMAN'S BAY AND MOUNT DESERT LAND AND WATER CO.,

By CHARLES H. LEWIS, President.

BEAUMONT STREET, DORCHESTER, MASS., September 4, 1889.

New England Adamant Co., Boston.

GENTLEMEN:—I have used your wall plaster **in three houses,** and am now, after a lapse of considerable time able to give you my opinion of its merits. There is in my judgment, no material yet before the public, which begins to compare with Adamant in all the important characteracteristics of a substantial wall.

First—It makes **the hardest** wall I have ever seen, and nothing but a blow with great force with some hard pointed instrument seems to make any impression upon it. When a hole is made by the driving of a nail into it, the hole remains the same, **not crumbling away,** and constantly increasing in size as is the case with lime plaster walls.

Second—It makes a beautiful **smooth wall**, perfectly adapted to painting or any other interior decoration.

Third—**The rapidity with which the walls dry,** and the consequent large saving of time in construction is another of its great advantages. I think it is by all means the only plaster to be used in schools, churches, halls, and all buildings which are exposed to hard wear and tear, by reason of its **great strength and durability.**

Finally, it is, in my judgment, the plaster of the future, as sure to supersede the common plaster wall as the modern railway has superseded the stage coach.

Very truly yours,   H. S. CARUTH.

Newport, R. I., April 15, 1889.

New England Adamant Co., Boston, Mass.

Gentlemen :—I am well pleased with the Adamant Wall Plaster. The disagreeable part of altering my house was the thoughts of waiting for plastering to dry. In using Adamant, I found that two days was amply sufficient to wait for the second coat, and the carpenters were able to put up the finish within twenty hours after it was put on. It dried out hard, white and smooth, and has been much admired by masons, carpenters, and one architect who called to see it. **I certainly should use it in either repairs, alterations, or new buildings.**

Yours respectfully,   T. MUMFORD SEABUBY.

West Newton, Mass, April 10, 1889.

New England Adamant Co., Boston, Mass.

Gentlemen :—We are **entirely satisfied** with the material you furnished for the repairs in our bank. I doubt if it would have been possible to use ordinary mortar where your preparation was used. We painted over it **in a few days,** and it was as dry and hard as need be. It seems to me to be quite an accession to the building fraternity's stock of material.   Very truly yours,

J. H. NICKERSON.
Treasurer W. Newton Savings Bank.

Bar Harbor, Me, April 18, 1889.

New England Adamant Co., Boston, Mass.

Gentlemen :—We have used the Adamant Wall Plaster on the walls of our new Congregational Church in this village, and find that it meets our expectations and your recommendations in regard to **economy** and **superiority** over the common lime plaster.

Very truly yours,   WILLIAM ROGERS,
Vice-President First National Bank.

HOLYOKE, MASS., April 15, 1889.

Mr. N. J. Brockway, Boston, Mass.

DEAR SIR:-- Yours of the 13th received, and would say that Adamant " fills my bill" for a plastering material, **I have used it entirely on houses for sale,** and if I promise a man a good job I shall use Adamant for plastering. Since using it I have come to the conclusion that good lime mortar is poor stuff.   Yours truly,

ED. NETHERWOOD.

WINCHESTER, N. H., May 29, 1889.

New England Adamant Co., Boston, Mass.

GENTLEMEN:--In reply to your inquiry as to how the material proves, we would say **that in every particular it has given perfect satisfaction,** and we think that the Adamant is the coming material for plastering.   Yours truly,

TAYLOR & BALL.

SPRINGFIELD, MASS., March 19, 1889.

New England Adamant Co., Boston, Mass.

GENTLEMEN:—I have used your Adamant Wall Plaster and believe it is **very much superior** to the lime plaster.

It is both hard and tough, and is **the best fire-proof material** that I have ever seen used inside of a house.

Yours truly,   JOHN McFETHERIES.

PEERLESS TOBACCO WORKS,

ROCHESTER, N. Y., August 14, 1888.

Bay State Adamant Co., Boston.

GENTLEMEN:—It was not without fear of disappointment that I decided to use Adamant in building a cottage at Nantucket, Mass., for workmen there are loth to experiment; however, I found one who was willing to make the attempt, though he had never seen the article used;

22

but the result is that he met with no difficulty whatever, **and was so well pleased** with it that he has taken the agency of it on the Island.

I can say that Adamant is all that is claimed for it, and should work a new era in house-plastering   Truly yours,

W. S. KIMBALL.

COTTAGE CITY, MASS., November 10, 1888.

Bay State Adamant Co., Boston, Mass.

GENTLEMEN:--In answer to your inquiry as to how I like the new plastering, I can say, with one exception, I am more than pleased, and that is not the fault of the Adamant, but the plasterer who put it on, he being a poor workman. Otherwise I think it all you recommend. I have had several gentlemen here to look at it, and probably you will receive orders from them. **It has been wet since it was finished, without at all affecting it.** We put white wood casings on within a week or ten days, without any apparent swelling. **It works splendid in patching up old plastering.** I shall give it a further trial in my home soon.   A. L. SAYLES.

WILTON, N. H., February 25, 1889.

Bay State Adamant Co., Boston, Mass.

GENTLEMEN:--I would say that your Adamant Wall Plaster is giving entire satisfaction. **In one week's time we can put on finish,** thereby saving from **four to six weeks** time over common mortar. I would especially recommend it **for patching and winter use.**

Yours truly,   H. L. EMERSON.

SHREWSBURY, MASS., January 7, 1889.

Mr. N. J. Brockway, Manager Bay State Adamant Co., Boston, Mass.

DEAR SIR:--Enclosed please find check in settlement of account. I cannot well use Adamant on my new house as the laths are so coarsely laid, and grounds are ¾ inch. I would give $25 if there was not a window-frame, door-jam, or lath on the house, for I like the Adamant walls very much, and so do my carpenters and masons. We found in one room which was lathed after I ordered, **it went fully as far as you claimed.** Every one who has seen the walls and ceilings I had put on like them very much, and I predict for it a very large sale. Had I known what I do now, I would willingly have given $50, for I think it would have been that saving to me.

Yours respectfully,   HIRAM W. LORING.

23

BOSTON, MASS., October 18, 1889.

New England Adamant Co., 21 *Federal St.*

GENTLEMEN:—We have used the Adamant Wall Plaster on **the new Tremont Theatre**, Boston, Mass., and heartily endorse all claims for it as stated in your circular.

We consider it **far superior** to any plaster ever put on the market for wire lathing.

We shall certainly use Adamant on all our work where a **first-class wall** is required. Yours respectfully, FRED A. GIDDINGS, Supt.

For Frank E. Smith, Builder, 150 Broadway, N. Y.

HOLYOKE, MASS., May 29, 1890.

N. E. Adamant Co., Boston, Mass.

GENTLEMEN:— With some hesitation, this being my first trial, I used your "Adamant Wall Plaster" in my new tenement block, containing 3800 square yards.

I watched its application with interest, and **am convinced** that **its merits**, as stated in your catalogue, **are not over estimated.** The plastering **is very hard**, and the **entire absence from cracks of any kind or description is wonderful.** I consider Adamant **far preferable** to lime plastering for the following reasons, viz: **hardness, adhesive qualities, no cracks** of any kind, in fact I consider that, in so far as it is possible, Adamant makes an indestructible wall.

The first cost, I find, is somewhat more than lime, but **with the above results** this is **no obstacle to me,** to say nothing about **the valuable time saved** by its use. You's truly,

WATSON ELY.

COCHESETT, MASS., May 5, 1890.

New England Adamant Co., Boston, Mass.

GENTLEMEN:— **One year and a half ago** our new church vestry was plastered with "Adamant." **Not a crack can be found upon the ceilings or walls.** Its smooth, marble whiteness remains in all its beauty. We have no wainscoting, but **the Adamant stands** all the chair thumpings **without crumbling or breaking.** It is undoubtedly the most perfect Wall Plaster known to the trade, in this age of the world. Yours truly,

R. J. KELLOGG, Pastor, Cochesett M. E. Church.

P. S.—I hear that our Baptist Brethren are equally well pleased with the "Adamant" walls on their new and handsome church, near here

R. J. K.

BOSTON, MASS., Oct. 30, 1889.

New England Adamant Co., 21 *Federal St.*

GENTLEMEN:—By the use of Adamant on my new block, School street, I saved at least one month's time in its completion and occupancy. I am also very much pleased with it in other respects.

Yours truly,　　　GEO. E. NILES.

BOSTON, MASS., Oct. 21, 1889.

New England Adamant Co., 21 *Federal St.*

GENTLEMEN:—We have used the Adamant Walls in the Boston College, 761 Harrison Ave., and find that they sustain, fully, every claim you make for them. **Their strength is phenomenal**, and for **beauty of finish** they are unsurpassed.

Weeks of time are saved in the completion of a building by their use. We cheerfully recommend them wherever **first-class walls** are desired.

Yours very truly,　　　T. J. FEALY,

Supt. of Building.

NEWTON UPPER FALLS, MASS., Jan. 28, 1890.

Bay State Adamant Co., 164 *Devonshire Street.*

GENTLEMEN:—Please find enclosed my check for $158.85 in settlement of enclosed bills for Adamant, "**Great Stuff**," and you can refer to me if you choose as one who has used it and who is a **firm believer** in its coming into **general use** for plastering.

Yours truly,　　　J. W. MITCHELL

BOSTON, MASS., Feb. 5, 1890.

N. J. Brockway, Esq., General Manager.

DEAR SIR:—After the recent fire in Boston, known as the Bedford and Chauncy street fire, in which my building was partially destroyed, rendering it necessary to replace all the plastering, it was suggested to me by my architect, Mr. C. H. Blackhall, that it would be a **large saving of time to use the material manufactured by your company and known as Adamant,** and acting upon his suggestion we plastered the walls and ceilings of the entire building with **Adamant, thereby saving,** in my opinion, **at least three weeks,** and while it has been on but a short time, so far **I am very much pleased with its appearance,** it being **much more even and clearer** than the ordinary plaster.

Yours truly,　　　WM. H. ALLEN.

FALL RIVER IRON WORKS CO.,

FALL RIVER, MASS., June 21 1890.

N. E. Adamant Co., Boston, Mass.

GENTLEMEN :—We used your Adamant Wall Plaster on wire lathing in our mill on the ceiling of our picker and dust rooms, also on stairway partitions for the following reasons, viz :

First—**The insurance underwriters strongly advised it.**

Second—As our mill was running it would have been **impossible to use anything else** without shutting down, causing a large loss in waste of time.

The Adamant put on at night was as hard as a rock in the morning. Heavy looms are running at a high rate of speed on the floor above the picker room, and the jar of the same has **not injured** the plastering in the least. **We are perfectly satisfied with the Adamant, as it has thoroughly answered the purpose for which it was used.**

EDWARD L. GRIFFIN, Treas.

WESTFIELD, MASS., Dec. 9, 1889.

New England Adamant Co.

GENTLEMEN :— I am happy to announce that the plastering in my house is finished, and is a **success in every particular.** From the time I commenced to think of building, the subject of plastering was a constant source of anxiety to me. I received your circular, and took plenty of time to investigate the subject of Adamant Wall Plaster, with the result I concluded it was just what I wanted. I expected something nice as a result of my investigations, but I am pleased to say that my expectations are more than realized. It seems to me to be the **only fit thing with which to plaster the walls of a house.** A great many people have been to see the new plaster, and the walls have been whacked and pounded enough to demolish a dozen walls plastered with lime, but there are no marks left to show the pounding. As stated above, it is all that I expected, and more; it is as **hard as a rock and of the most beautiful finish.** You are at liberty to refer to me in this matter. Yours truly, WM. H. JOHNSON,

Of Johnson & Son, Church Organ Builders.

WORCESTER, MASS., April 16, 1890.

New England Adamant Co., 21 Federal St., Boston.

GENTLEMEN :—The " Moen " house ot this city, which I plastered with " Adamant," is the finest piece of workmanship I have ever seen. It is simply perfect. It has fully demonstrated that " Adamant Walls " are

26

all, and more than you claim for them. **They are as true and smooth as polished marble, and almost, if not quite as hard.** All of the ceilings are plastered on wire lath, of ordinary mesh, ($2\frac{1}{2}$ to the inch) were applied without the least trouble or waste of material, and **are as hard as stone.** By the use of Adamant on this house I am able to complete it **a full month earlier** than if I had used lime. There is **not a crack or blemish in the whole job.** I believe Mr. Moen's house to be the best job of plastering in the city of Worcester. You are always at liberty to refer to me as to value of your walls, to which I give my full endorsement. Yours very truly,

JOS. G. VAUDREUIL,

Contractor and Builder.

NORTHAMPTON, MASS., March 14, 1890.

New England Adamant Co., Boston, Mass.

GENTLEMEN:—I plastered the First National Bank Building of this city with your Adamant Wall Plaster and am very much pleased with the results. The walls are very hard, smooth, and of a most beautiful finish. Saved fully three weeks' time in the completion of my work on account of its quick drying qualities. **Had occasion to remove a small portion of the plastering and found it adhered firmly to the laths, not being dependent on the clinch to hold on the wall, thus adding great strength to the building.** Cut holes through the plastering for steam pipes, with a saw, and did not break the edges or crumble them, but cut as clean and smooth as a board could be. I consider Adamant in every particular vastly superior to lime mortar for plastering. Yours truly, J. L. MATHER.

SPRINGFIELD, MASS., May 27, 1890.

N. E. Adamant Co., Boston, Mass.

GENTLEMEN:— The contract for plastering my new hotel had been awarded common lime mortar, but before the plastering was started, my attention was called to your material, and after a careful examination of your claims, I took counsel with my architect, Mr. Jason Perkins, and used Adamant Wall Plaster throughout my entire hotel addition, about 12,000 square yards, also used large quantities in repairing my old building. My experience is that Adamant is THE plastering for hotels. It is true to its name in hardness. **There is not a chip, crack, a lath crack, or a pit in the entire building.**

Aside from its actual value, which alone should cause it to be **univer-**

sally used, I opened my hotel at least one month earlier than I possibly could have done had I used lime mortar. **This saving of time alone makes Adamant a profitable investment.**

I am perfectly satisfied with your walls, as all of **your claims have been verified** by my experience. Yours truly,

J. M. COOLEY.

From Hon. Byron Weston, ex-Lieut. Gov. of Massachusetts.

DALTON, MASS., Jan. 14, 1890.

New England Adamant Co., Boston, Mass.

GENTLEMEN:—Mr. Dodge plastered my "Riverside Block" in December with the Adamant Plaster because it could be done **safely in cold weather.**

It became set in a few hours and **hard as marble.**

Had no trouble about freezing or drying out.

It was so hard that the workmen or carpenters **could not dent or scratch it.**

I have a **perfect** hard finish wall.

I see many benefits over a lime mortar wall.

Yours truly,

BYRON WESTON.

CHICOPEE FALLS, MASS., May 28, 1890.

N. E. Adamant Co., Boston, Mass.

GENTLEMEN:— In my new block in this city I used your Adamant Wall Plaster throughout the stores on the first floor. My contractor, Mr. F. F. O'Neil of Holyoke, also my architect, Mr. G. P. B. Alderman, urged me to use it on the tenements above, but as I knew nothing of its merits, did not care to pay $400. extra, so used lime mortar.

Your claims have been verified, and I would gladly give the $400. could I now have the Adamant substituted for the lime mortar on the tenements. I am about to build another large block and **shall use your plastering throughout.** Yours truly,

DANIEL DUNN.

WORCESTER, MASS., May 22, 1890

New England Adamant Co., Boston, Mass.

GENTLEMEN:—My new house on Elm St. is plastered with your "Adamant Wall Plaster," and it gives me **complete satisfaction.** The walls are of a **beautiful finish,** they have **great strength** and have **saved me valuable time** by their quick hardening and drying qualities. I believe your walls are **in every way superior** to walls of lime and sand, and give them my cordial endorsement.

Yours truly, PHILIP W. MOEN.

28

OFFICE OF O. J. LEWIS, BOOTS AND SHOES, }
68 LINCOLN STREET,
BOSTON, MASS., June 5, 1889. }

New England Adamant Co.

GENTLEMEN:—Yours asking how I liked the Adamant Plaster duly received. **I reply by enclosing an order for enough to put on my cottage.** Yours, O. J. LEWIS.

SPRINGFIELD, MASS., October 15, 1888.

Bay State Adamant Co.

GENTLEMEN '—I forward enclosed check which please acknowledge. The work, I think, is entirely satisfactory to all concerned, **and my workmen can handle the material as readily as ordinary plaster.** Will be glad to assist you to any further work in this vicinity.

Respectfully, J. S. SANDERSON,
Contractor and Builder.

SORRENTO, Me., Aug. 1, 1888.

Bay State Adamant Co , Boston, Mass.

GENTLEMEN:—I shipped your empty bags by B. & B. S. S. Co. Please excuse the delay.

I am very much pleased with the plastering. A great many people speak of **the fine finish** in our parlor and office, where No. 1XXX was used.

Give me your best prices, and send me a catalogue, so that I may know what to tell people who are building.

Respectfully yours, W. H. LAWRENCE.

PORTSMOUTH, N. H., June 21, 1889.

New England Adamant Co., Boston, Mass.

GENTLEMEN:—The Adamant Wall Plaster which you put on the rooms in the tower of my brewery proves **very satisfactory.** It hardened up well and makes **very smooth, clean surface.**

Very truly, FRANK JONES.

ELLSWORTH, Me., June 17, 1889.

New England Adamant Co., Boston, Mass.

GENTLEMEN:—Yours of 11th inst. at hand. In reply would say that I used your plaster for a ceiling which took about thirty barrels. After it was finished there came a heavy rain and the roof leaked very badly.

When I discovered it there was about **three inches of water** all over the ceiling, and it remained there two days, very little of it soaking through. After it dried I put on a coat of kalsomine to cover stains, and I think it looks as well as it did before being wet. I cannot see that it has started any from the laths. **If it had been plastered with lime mortar it would have come off in one day** with that amount of water on it. I think that Adamant is all you recommend it to be and will soon take the place of common plaster.

Respectfully yours,   P. H. STRATTON, Contractor and Builder.

SPRINGFIELD, Mass., June 6, 1889.

New England Adamant Co., Boston, Mass.

GENTLEMEN :--Enclosed please find check for $404.67, in payment for Adamant used on the French Protestant College. I shall induce my customers to use Adamant whenever I can, as **I believe it to be the best** in many respects, **of any plastering ever introduced.** My house is about half plastered, and I will remit soon for Adamant used.

Yours very truly,   S. E. WALTON.

BAR HARBOR, Me., April 9, 1889.

Bay State Adamant Co., Boston, Mass.

GENTLEMEN:--I think for good work the Adamant is preferable to common plaster, especially when used in freezing weather, as it becomes **hard and solid** all through in a few days.

The Adamant on the church, I am pleased to say, has proved all that was claimed for it.   Yours truly,   ASA HODGKINS, Contractor and Builder.

GARDNER, Mass., April 16, 1889.

New England Adamant Co., Boston, Mass.

GENTLEMEN:—Yours of 13th inst. received, and in reply would say that I used Adamant on an old house last season over old-fashioned split board lath, where many of the spaces were so narrow that common mortar would not hold well. **Adamant worked well,** though expensive, as many spaces were wide, but **have a splendid job.** For patching up old work I **never saw its equal,** as it sticks to the old mortar, makes a solid joint and dries so quickly that it saves much time and trouble. Two rooms that I was living in I cleaned out Wednesday, took off old mortar and first-coated Thursday, second-coated the following Monday, and Tuesday had painters and paper-hangers at work on them. It is also **much more convenient** to mix Adamant on the floor where you are at work, particularly in wet weather, where with common mortar we should be obliged to suspend operations during storms. The job I did with it was not perhaps a fair test, so could not say what the extra expense would be on new work where lathing was intended for Adamant, but should judge it would cost no more than a good three coat in lime mortar.

Yours in haste,   C. F. BOUTELLE.

WINCHESTER N. H., April 15, 1889.

New England Adamant Co., Boston, Mass.

GENTLEMEN:—I think the Adamant Wall Plaster that I purchased of you is **the best** I ever saw. I would not put on lime mortar if it was given me, providing I could get Adamant. I think it **cost me less** to plaster my house with Adamant than it would had I used lime mortar, as I did not keep any fires to dry it as I should have had to do with mortar. There has been a number of people to see my walls. The Universalist Church of this place is plastered with Adamant and it makes nice walls.

Respectfully yours,   CHAS. D. SEAVER.

S. A. SWEETLAND, CONTRACTOR AND BUILDER, }
NATICK, Mass., Jan. 26, 1888. }

Bay State Adamant Co., N. J. Brockway, Supt.

DEAR SIR:—I take great pleasure in recommending your Adamant Wall Plaster. I have used it at Wellesley College, and I acknowledge it a great success. **As to its hardness, quick drying, and complete surface, it has no equal, and it is a pleasure to put up the woodwork finish,** for there is no danger of nicking or cracking, as it is almost **impossible to break it with a hammer; and no patching** for the masons after it is up.

Yours truly,   S. A. SWEETLAND,
                              Contractor and Builder.

MARLOW, N. H., May 4, 1889.

New England Adamant Co., Boston, Mass.

GENTLEMEN:—In reply to your inquiry as to how we liked the Adamant Plaster, would say we used it the first week in January on kitchen and pantry. **It dried very rapidly and without a crack.** The next week the carpenter put in pantry shelves and the pounding **had no effect** on the plastering whatever.

Yours truly,   A. E. FLAGG.

31

The following is a copy of a private letter from Mr. Russell to a personal friend:

SYRACUSE, N. Y., May 16, 1889.

Warren F. Draper, Esq., Andover, Mass.

MY DEAR SIR:—Your letter of inquiry relative to the Adamant Wall Plaster came to me this morning. In reply I would say that I have used this material referred to in various classes of buildings for the past two years and **always with the most satisfactory results.**

It is true, that through inferior workmanship, negligence or careless-ness, bad work may result even with the best of any materials, but in the use of Adamant there should be no fear of failure either in **hardness, durability, facility of application or beauty of finish.**

The Adamant walls of the John Crouse College of Syracuse Univer-sity, are eminently satisfactory. For this class of buildings and for your bui ding at Andover, as well as others **where good plastering is desired,** Adamant cannot be surpassed

With workmen who are familiar with its use, it is susceptible to dec-orative treatment, both varied and interesting, and its surface can be finished to receive mural decorations in fresco, paper or other methods most successfully.

Its hardness is correctly expressed in its title. You can make no mistake in its use.

I give you the above opinion as the result of **my own practical ex-perience as an architect,** and from no superficial or heresay inform-ation. Very truly yours, ARCHIMEDES RUSSELL, Architect.

MINNEAPOLIS, MINN., June 14, 1889.

N. W. Adamant Manufacturing Co.

GENTLEMEN:—I have used some 7,000 or 8,000 yards of your material on a building for which it was found it was difficult to get good local materials. The base coat of Adamant is very strong and the finish work very strong and hard. The mechanical mixing of the materials in defi-nite proportions is capable of putting the plasterer's work on a much surer basis than anything we have had heretofore. It goes without saying that a material which combines with the mixing water, chemic-ally, instead of allowing it to evaporate slowly and be in such large part absorbed by the timbers and which will not " pop " out will commend itself to the building public. Yours, etc.,

F. G. CORSER, Architect,

402 Nicolette Ave.

MINNEAPOLIS, MINN., June 20, 1889.

N. W. Adamant Manufacturing Co.

GENTLEMEN:—From what I have seen of the Adamant, I think your claims are well taken, and that you have the very best material in way of plaster that has been put on the market. It makes a very hard, tough wall that will stand wear and keep out cold, moisture, and sound, I shall be glad to use it whenever I can do so, feeling that the "best is the cheapest." Very truly yours,

WM. CHANNING WHITNEY, Architect,

Rooms 34 and 35 Tribune Building.

CLEVELAND, O., May 16, 1889.

The Detroit Adamant Co.

GENTLEMEN:—I am satisfied that Adamant is the coming Wall Plaster, and am willing to testify to its merits.

Yours very truly,   E. P. RUPRECT, Architect.

MINNEAPOLIS, MINN., April 25, 1889.

N. W. Adamant Manufacturing Co.

GENTLEMEN:—It is with pleasure that I endorse your many claims for Adamant Wall Plaster. Its hardness and elasticity are marvelous when compared with common lime plaster. Its advantages in quickness of setting permit the carpenters to suffer no delay. I most cheerfully recommend it to the public, and have confidence in its future.

Yours very truly,   L. S. BUFFINGTON, Architect,

Boston Block.

OFFICE OF J. H. KIRBY, Architect,

No. 18 LARNED BUILDING, SYRACUSE, N. Y., May 25, 1887.

The Adamant Wall Plaster Co. Syracuse, N. Y.

GENTLEMEN:—It gives me pleasure to recommend your Adamant Wall Plaster. I have used it on several buildings and find it possesses all the good qualities claimed for it, namely: Strength, toughness, adhesiveness, and general durability. I think it is the best wall plastering I have ever used on any building, and most heartily approve of and endorse the Adamant Wall Plastering.

Yours respectfully,   J. H. KIRBY.

33

CONSTABLE BROS. & T. MELLON ROGERS,
Associate Engineers and Architects,
902 Walnut St., Philadelphia, Pa., and 149 Broadway, N. Y.

PHILADELPHIA, PA., November 24, 1888.

The Adamant Plaster Co., 210 S. Tenth Street, Philadelphia.

It gives us great pleasure to testify to the great worth of your Adamant Plaster, and to assure you that in every case where we have used it of its having given entire satisfaction, and in all cases of winter work especially is it valuable, on account of its non-freezing character, and its always saving from three to six weeks' work; we intend using it in every building that we possibly can, it being so much more reliable and better in every way than common plaster.

Respectfully yours, CONSTABLE BROS. & T. M. ROGERS.

NEWARK, N. J., May 8, 1888.

The New Jersey Adamant Manufacturing Co.

GENTLEMEN:—I can cheerfully recommend your Adamant Plaster to all who may require quick and serviceable work. I have used it in the Clinton Building, Newark, and it has given entire satisfaction. I will be pleased to give any one an opportunity of examining it on the wall of the said building, and you may refer to me at any time.

Yours respectfully, HENRY D. HAVILL, Architect.

JOLIET, ILL., November 3, 1887.

Adamant Manufacturing Co., Syracuse, N. Y.

GENTLEMEN:—Replying to yours of the 1st inst., will say that I have used Adamant Wall Plaster on the new Union School in this city, and am well pleased with it. I can cheerfully say that it does all you claim for it, and that I consider it the only strictly first-class wall plaster.

Respectfully yours, F. S. ALLEN, Architect.

PHILADELPHIA, PA., November 25, 1887.

Keystone Plaster Co.

GENTLEMEN:—We are using Adamant Wall Plaster on all of our work. It not only gives entire satisfaction, but it saves us from three to four weeks in the construction of an ordinary house. We are willing always to recommend it and give our reasons for doing so to all who consult us on the subject.

Yours truly, CULVER & ROGERS, Architects,
901 Walnut Street.

SYRACUSE, N. Y., December 6, 1886.

This is to certify that I have used the Adamant Plaster throughout one of the best houses in this city, and that I am very much pleased with it. The result is more satisfactory than any hair mortar can be. I shall use it in the future when available.

Very truly,　　　　C. E. COLTON, Architect.

PITTSBURGH, PA., November 2, 1888.

Keystone Plaster Co.

GENTLEMEN:—I am happy to say that a trial of your Adamant Wall Plaster has convinced me that it fully justifies your claims for it, and I cheerfully recommend its use.

Very truly yours,　　C. M. BARTBERGER, Architect,

Room 62 Lewis Block.

---

751 BROAD STREET, NEWARK, N. J., May 8, 1888.

The New Jersey Adamant Manufacturing Co.

GENTLEMEN:—I have used the Adamant Wall Plaster in the buildings of the "New Jersey Home for Disabled Soldiers," in Kearney, N. J., and the results are very satisfacto y As it is greatly superior to the old methods of plastering. I intend using the Adamant in other buildings hereafter.　　　Respectfully yours,

PAUL E. BOTTICHER, Architect.

SYRACUSE, N. Y., December 7, 1887.

Adamant Manufacturing Co., Syracuse, N. Y.

GENTLEMEN:—Your request of the 2d duly at hand. Please excuse delay in replying. We are perfectly satisfied with the results obtained by the use of your Adamant Wall Plaster. Among many advantages which we see in its use over the ordinary lime mortar are the following: No checking or cracking in drying; no "pitting;" a great shortening of the usual delay in the carpenters' work by the quick setting and drying of the walls; its durability arising from its extreme hardness and resistance to knocks, and also its water-proof qualities. We have specified the use of your material for about 20,000 yards during the season now closing, and we feel perfect confidence in using your material, and would refer the inquiring public especially to your work in "The Florence Building," where Adamant was used.

We are, Yours very truly, BAXTER & BUELL, Architects,

Nos. 4 and 6 Butler Block.

35

N. W. Adamant Manufacturing Co.

GENTLEMEN:—I take pleasure in testifying in behalf of your Adamant Wall Plaster. Through my own experience I can freely say it is the best material for wall plaster so far brought to my notice, and I am satisfied that it possesses all the advantages and qualities you claim for it.

Yours truly,   GEO. E. BERTRAND, Architect,

810 Lumber Exchange.

GEORGE C. MOSER, Architect,

NORFOLK, VA., May 22, 1889.

The United Adamant Plaster Co., Baltimore, Md.

GENTLEMEN:—Yours of the 21st received; noted. In answer to your inquiry I cheerfully say, that after a thorough examination of the materials, on and off the wall, I like Adamant; in fact, so much was I impressed with its excellence and superiority over any other plaster or wall finish, that I at once specified it for a block of four stores, then out for estimates, and will continue to use it in my work.

Yours, &c.,   GEO. C. MOSER.

OFFICE OF M. H. HUBBARD, ARCHITECT,   }
UTICA, N. Y., June 12, 1889.   {

Adamant Plaster Co.

GENTLEMEN:—I have been specifying your Adamant Plaster for nearly two years, and I consider it the coming plaster. It is far superior to any plaster I have ever used, especially for churches, of which I am making a specialty. I used it on the Oneida Baptist Church, and I find it is a grand thing for the acoustic properties, and in every respect it is far more than recommended to be. I am placing it in my specifications as a permanent fixture.   Respectfully,   M. L HUBBARD.

409 Bank of Commerce Building.

MINNEAPOLIS, MINN., June 14, 1889.

N. W. Adamant Manufacturing Co.

GENTLEMEN:—After having given your Adamant Plaster a thorough inspection and test, I am using it on all the buildings erected by me this year, both in this city and out of town, and can say that it has given me the very best of satisfaction.

Yours,   EDW. STEBBINS, Architect,

173 GENESEE ST., UTICA, N. Y., January 14, 1888.

Mr. James S. Foresman.

DEAR SIR:—In reply to your favor of the 12th, I will say that my experience with Adamant Plaster has convinced me that either for public or private use it is the coming material. I have used it in both classes of work and if possible would use no other material for plastering. Having all the advantages of quick lime mortar with none of its defects, I see no reason why it should give even other than entire satisfaction.

Yours respectfully,   G. EDW. COOPER, Architect.

NEW YORK, June 11, 1890.

Adamant Manufacturing Co., Bennett Building.

GENTLEMEN:—Your "Adamant Wall Plaster," which has been on my house at the southeast corner of Pierrepont and Henry Sts., Brooklyn, for over a year, is so satisfactory that I and all whose attention I have called to it are of the opinion that it is superior in every particular not only to common lime and sand mortar, but all other plasters we have ever seen. I deem myself fortunate in having used it. It is as hard as a rock and shows no cracks. Yours very truly, HERMAN BEHR.

PIERREPONT, N. Y., March 3, 1890.

GENTLEMEN:—I used the Adamant Plaster for my house last year and am much pleased with it, as it makes a very hard wall that does not crack or crumble when casings are being put on, and also **makes a house warm and free from dampness.** I think it pays for all the extra expense, and **no good house is well built without it.** The acoustic properties of rooms are very fine, and music has a much better effect than in rooms plastered with the ordinary material. You may refer customers to me at any time. . Yours truly,

W. H. GRENELL.

MINNEAPOLIS, MINN., June 22, 1889.

N. W. Adamant Manufacturing Co.

GENTLEMEN:—I have used your Adamant Wall Plaster and I think I have the best plastered walls in this State. It is the coming plaster, and in time no other will be used.

Yours truly,   A. L. DORR,
507 Lumber Exchange.

OFFICE OF TRUAIR & WYATT, Insurance Agents,
SYRACUSE, N. Y., December 12, 1888.

GENTLEMEN:—My house was plastered with your goods last spring, and I can certify to the fact that I have an A No 1 wall. Adamant is a great success, and has given me the utmost satisfaction. I consider my house is at the least worth $500 more than if common plaster had been used. Respectfully, GEO. W. WYATT.

A. PASQUINI, Mason and Builder,
No. 208 SHERMAN ST., ALBANY, N. Y., Dec. 14, 1888.

Adamant Manufacturing Co.

GENTLEMEN:—I have used the Adamant Plaster on the Albany County Bank, and was very much pleased with it. I have since used it on several buildings, and find it superior to the ordinary lime and sand mortar. It has given entire satisfaction in each instance by its strength and hardness. It is very easy to work, saves labor and time, there being no breakage or cracking in drying out. I cordially recommend it to every person who wishes a first-class wall plaster.

Respectfully yours, A. PASQUINI.

OFFICE OF ARKELL & SMITHS, Flour Sack Manufacturers,
CANAJOHARIE, N. Y., May 31, 1887.

The Adamant Manufacturing Co., Syracuse, N. Y.

GENTLEMEN:--In answer to your inquiry of May 28th, with regard to the Adamant Plaster put upon my house, would say that it is pre-eminently satisfactory, and yet was put upon the building under peculiar circumstances. I had my house already plastered with ordinary lime plaster, when (on account of putting down double flooring) was so broken and jarred by the hammering as to come down in large quantities, and was obliged to take it all off, and then to put on the material that you manufacture. This was done without affecting any of the wood-work, which was already up and finished. Were able to go on with the work without any detriment from moisture, and can satisfactorily endorse your claim as to its being a complete and efficient material for the purpose for which it is intended. Wishing you the success that your enterprise deserves, I am,

Very truly yours, WILLIAM J. ARKELL.

38

WILLIAMSPORT, PA., September 12, 1888.

Keystone Plaster Co.

GENTLEMEN :—In the re-construction of a portion of our seminary buildings we covered more than 7,000 yards of walls and ceilings with your Adamant Plaster. The result has been entirely satisfactory. The Adamant is especially well adapted to schools, hotels, public halls, and all buildings which are much used, as it is extremely hard, and at the same time sufficiently pliable not to crack or break.

Very respectfully,   EDW. J. GREY, D.D.,

Prest. Dickinson Seminary.

NEWARK, N. J., May 7, 1888.

New Jersey Adamant Mfg. Co., Harrison, N. J.

GENTLEMEN :—I take pleasure in recommending your Adamant Wall Plaster. I have used it at the New Jersey Home for Disabled Soldiers, Kearney, N. J., and I acknowledge it a great success. As to its hardness, quick drying, and complete surface it has no equal, and it is a pleasure to put up the wood-work finish, for there is no danger of nicking or cracking. It is almost impossible to break it with a hammer, and no patching for the masons after it is up. I would not use any other plaster in building for myself, and would recommend it to every one building.   Yours truly,   E. B. VLEIT,

Carpenter and Builder.

SYRACUSE, N. Y., May 19, 1887.

The Adamant Mfg. Co., Syracuse, N. Y.

DEAR SIRS :--The Adamant Wall Plaster which was used throughout my house has given me great satisfaction. There are many reasons why I think it superior to the common mortar. It is ready to be used whenever needed. It dries so quickly that it does not swell the wood, thus enabling the carpenters to almost immediately follow the masons. It is so hard it does not bruise while casings are being put on, nor is it marred by furniture. It resists the action of water. While the house was building the tank overflowed, and the kitchen ceiling was flooded. Holes were bored through the ceiling, a large quantity of water was drawn off, and when the plaster dried we found that it was not injured in the least, save the discoloration caused by water. I should certainly use it if I were to build another house.

Yours truly,   S. T. FORD,

Pastor Central Baptist Church.

Adamant Manufacturing Co.

GENTS:—Last summer I had built for myself a house, and, upon the representation of the builder was induced to use the Adamant Plaster manufactured by you. To say that I am pleased with it but faintly expresses the idea I wish to convey. I can honestly and cheerfully recommend it those contemplating building, and believe they will never regret using the Adamant instead of the old-fashioned mortar. Its superiority over the latter is easily demonstrated: 1st—In an almost entire absence of cracks. In the ten rooms of my house I cannot find one crack in a ceiling or side walls, and only in the angles do they appear, and even there they cannot be charged to the account of the plaster, but rather to the shrinkage of the timber, some of which was not too well seasoned. 2d—Its hardness and durability are strong points in its favor, and a person will fully make up for the small extra cost in these points alone, to say nothing of others. Last, but not least, through the severe spells of cold weather which we have passed I have experienced no difficulty whatever in keeping my house at a comfortable temperature day and night, and in no part of the building, from cellar to garret, has frost penetrated. I attribute this in a great measure to your plaster. the hard, stonelike walls seeming to be impervious to cold. I could name other points of excellence, but these will suffice, and will conclude by saying that the Adamant is in every respect what it was represented to me to be, and I can cheerfully endorse anything you have claimed for it.   Respectfully yours,   W. R. CHAPPLE.

St. Meinrad's Abbey,

ST. MEINRAD, IND., August 30, 1888.

Keystone Plaster Co.

GENTLEMEN :— We are highly pleased with the Adamant Wall Plaster furnished us for our new college buildings. It is certainly much superior to any plastering material I have ever seen, and it is the only material fit to be used on schools and public buildings.

Yours truly,   FINTAN MUNDWILER, Abbott.

NEW HAVEN, CONN., March 28, 1889.

The Connecticut Adamant Plaster Co.

GENTLEMEN :— About two years ago I had occasion to rebuild a portion of my residence, and was induced to use your Adamant Plaster in finishing the walls. I had but little faith in your "New Material," and ried it against my judgment. I now desire to say that after this experience I am more than satisfied with it. Your recommendation has been more than realized. The walls are smooth and hard, without a crack or flaw. There is no crumbling if a hole is made in it for a picture knob or for any other purpose, and the e is no bruise if a chair back rests against it. Any ordinary blow, such as is often made by children, does not deface it. Dampness does not affect it, and stains are readily removed without injury to it. In every respect it has exceeded my expectations and your recommendations, and I can most heartily endorse all its friends say in its favor. Very truly yours,

LUCIUS P. DEMING, Attorney-at-Law.

UTICA. N. Y., December 13, 1888.

Adamant Manufacturing Co.

DEAR SIRS :—I used your Adamant Wall Plaster a year ago at the State Lunatic Asylum, Utica. I plastered about eight thousand yards, and it gave perfect satisfaction. It proved to be the best plaster that ever was put on any wall. I have also used this material many times since, and have recently done more work with Adamant at the Asylum. I heartily recommend it to every one who wants to plaster a wall. Hoping that your Adamant will take the place of all this common mortar, I remain, Yours respectfully,

PIUS KERUER,
Builder and Contractor.

SYRACUSE, N. Y., November 3, 1887.

To the Adamant Mfg Co., Syracuse, N. Y.

GENTLEMEN :—The plaster you put on our observatory, both directly on the brick work and on the lath, has, up to this time, given us perfect satisfaction. Its quickness of setting, hardness and neatness all indicate its great superiority over the old-fashioned plaster. I have the most perfect confidence in it as altogether the best plaster I have ever seen.

Yours truly, C. N. SIMS,
Chancellor of Syracuse University.

OGDENSBURG, ALEXANDRIA BAY, N. Y., Jan. 24, 1889.

Adamant Manufacturing Co., Syracuse, N. Y.

GENTLEMEN :—I reply to your inquiry as to how I was pleased with the Adamant I used for plastering Mr George M. Pullman's house at

Alexandria Bay, "Castle Rest," by saying that Adamant fills a long felt want, is a boon to builders when pressed for time, which I appreciate then and now. Makes a wall unexcelled for strength and hardness, and is the equal of any in beauty of finish. Yours truly,

S. G. POPE, Contractor and Builder.

Superintendent's Office, PHILADELPHIA, May 25, 1888.
Department of Charities and Corrections, Geo. Roney, Supt.

To the Keystone Plaster Co.

GENTLEMEN:— The introduction of your Adamant Plaster into the Blakely Almshouse some eighteen months or two years ago, appears to have opened up a new era. The abolition of large mortar beds, which are an eyesore of themselves, outside of many other considerations, justify me in thus making public my high appreciation of said plaster. For hospitals or large institutions it is to my mind the only thing that will meet the requirement of the present day. I will explain personally to all comers my many reasons for thus recommending the Adamant Wall Plaster. This you can use as you see proper.

Respectfully yours,   GEORGE RONEY, Superintendent.

BERWYN, PA., March 8, 1889.

Newton H. Culver,

Secretary United Adamant Plaster Co., Baltimore, Md.

MY DEAR SIR:—It affords me great pleasure to bear testimony to the almost numberless advantages your "Adamant" holds over ordinary plaster, and more especially from a sanitary standpoint is its use bound to become general. Hygienists have for years been looking for just such an article, for by reason of its chemical composition, which returns it as near as possible to the original rock or gypsum, it renders the wall to which it is applied, absolutely impervious to disease germs. It will therefore, not only in private residences, but more especially in hospital buildings, be found the wall plaster "par excellence" in the future.

Very truly yours,   T. L. ADAMS, M. D.

A. W. Palmer, Leading Clothier,

15 & 17 S. Salina St.,

Syracuse, N. Y., May 20, 1887.

Adamant Mfg. Co., Syracuse, N. Y.

Gentlemen :—Replying to your question whether I am pleased with the Adamant plastering on my house, I would say that I am. The strength of the wall is most remarkable. It has been hammered and pounded without damaging it in the least, in a manner that, in my opinion, would have utterly ruined any wall I ever saw made of common mortar. The facts that the Adamant became hard in a single day after being applied, and that frost had no effect upon it afterward, were important to me, because my house was built in the winter, and with common plastering I should have had to wait weeks, and to burn many tons of coal before the carpenters could go on ; but as it was they worked in one part of the house while the masons were working in another. Again, in putting on the casings, the plaster was not broken or injured in the least, and consequently no repairing was required. The saving in this item and in coal, will, I think, more than offset the difference in the cost per yard, and I have a wall vastly superior to one made of common mortar. Yours truly,

A. W. PALMER.

New York, June 10, 1890.

Adamant Manufacturing Co.

Dear Sirs :— It gives me pleasure to state that I am not only highly pleased with the Adamant Plaster you furnished me, but that it has **saved me over one hundred dollars per each house**, and made an **excellent and fire-proof job.** I shall continue to use it right along.

Yours respectfully, FRANK E. SMITH.

Syracuse, N. Y., Nov. 18, 1889.

Adamant Manufacturing Co., Syracuse, N. Y.

Gentlemen :— I desire to express to you my satisfaction with the Adamant Wall Plaster which I used on the Florence Apartment Building, about two years ago. I am so well pleased with its strength, durability and finish, that **I have just entered into contract for its use on my large block now in process of construction.**

Yours truly, J. M. WRIGHT.

43

Office of Adrian G. Hegeman & Co., Real Estate, 1321 Broadway,
NEW YORK, June 9, 1890.

Adamant Manufacturing Co.

GENTLEMEN:— The basement wall in the 34th St. house which has
caused so much trouble for the past ten years from dampness, has
been perfectly dry since you put on your Adamant Wall Plaster—
October 1889—and we recommend your plaster as being in every
way satisfactory.　　Yours truly,

ADRIAN G. HEGEMAN & CO.

SYRACUSE, N. Y., May 19, 1887.

Adamant Manufacturing Co.

Having used the Adamant Wall Plaster on several houses built under
my supervision the past year, I can recommend it as being superior to
all other plaster in the following respects: 1st. It makes a hard,
smooth wall, and does not check or crack in drying. 2d. There is no
pitting or crumbling, as is often the case with ordinary plaster. 3d. It
can be mixed for use in the room where wanted, and any plasterer can
put it on more easily and rapidly than the common mortar. 4th. So
much of the moisture is crystalized in the material that buildings are
not subject to the swelling and shrinking which ordinary plaster always
occasions. 5th. There need be no delay in waiting for walls to dry, as
the woodwork can be put up in a few hours after the Adamant is put
on. 6th. In finishing buildings this year I have been able to prepare
them for occupancy one month earlier or quicker than I could have done
by using any other plaster. 7th. In winter I have found the Adamant
would dry in a few hours, where ordinary plaster would have required
days and a good fire to produce similar results. 8th. Every person for
whom I have used Adamant has been better pleased with his house or
building than he would have been had it not been used.

ISAAC SMITH, Contractor and Builder,

No. 6 Raynor Avenue.

(From The Northwestern Builder and Decorator, March, 1890.)

A Factor in Heating and Ventilation.

If we may be permitted to make a few dogmatic, or axiomatic, if you please, assertions, the discussion of the problem will be simplified. There is but one kind of heat. The law of gravity is the same to-day as yesterday. The position for foul air exits is at or near the floor. From two to three thousand cubic feet of air per hour for each person, occupying a room are needed for health and comfort.

Because of the law of gravity, warm air will rise to the top of the room, and to remove it through the exits at the floor, force is required, the amount of which is to be determined by certain conditions existing in and peculiar to each case. To avoid drafts, and more particularly to prevent the accumulation of stagnant air in certain parts, generally corners of a room,—a common fault of school-room ventilating systems—each room should have several foul air exits, and the best places for them are at the cornes of a room, thus drawing the air from the upper part of the room to the floor in one uniform stratum rather than in a current, which is always caused by a single exit. This re quirement of the conditions present in almost every room and of the conditions fixed by the amount of air essential for health and comfort, suggests that the proper place for the entrance of the warm air is near the ceiling, for if placed in the floor an approximately uniform temperature in all parts of the room is not obtainable, for the ascending current causes a cyclonic movement of all the air in the room, and interferes with the system of ventilation.

With these preliminary remarks, of which the truth seems self-evident, we can approach the subject proper of this article, namely, a factor in heating and ventilation, which deals with the greatest difficulty in the way of the heating engineer—a factor almost, if not totally, ignored by every writer with whose writings we are familiar, resulting in either partial or complete failure to attain results sought for and admitted to be essential to a perfect system, using the word perfect only in a general sense and as descriptive of that which is possible of attainment, in a way, however, not practicable under ordinary conditions

In the summary of the conditions met with in the Metal Worker problem, the writer, an expert, gives cereain data of which every intelligent heating engineer knows the value. Among the data are the amount of glass surface and the amount of exposed wall surface and the direction of its exposure, in each room. It is the latter point that furnishes the text of this article. If we are not mistaken, not one of the writers even referred to the character of the exposed wall, and we may rightfully infer the same indifference existed as to the character of the ceiling, and as little or no heat was supplied to maintain a proper and uniform current in the exhaust flues, the problem resolved itself into this—an attempt to admit warm air, at one side near the bottom, into an inverted seive and to distribute it uniformly to all parts of the seive, and finally coax it out of a hole on the other side and at the bottom. Notwithstanding the absurdity of this comparison, it is almost literally true, for it has been shown time and again that common lime mortar is simply a seive, through which air will pass whenever its equilibrium is disturbed sufficiently to cause a gentle movement among its particles. Such disturbance is almost constant, either from heat applied within the room or from wind, a result of outside heat.

The writer has long made a study of building materials to determine the best and cheapest way of making a house warm at the least expense. Putting aside the more expensive non-heat conductors, such as mineral wool, etc., all builders, we think,, in this climate have adopted paper, but a single thickness of a poor and cheap quality, put onto the siding with carelessness, furnishes a very slight wind-break, while its absence from the ceiling is quite universal. Back plaster composed of lime and mortar is so nearly useless it would be expensive at a tenth of the cost of putting it on, and for reasons to be given hereafter. By the use of matched siding, inside and out, covered with two or even three thicknesses of the best quality of building paper, the outside walls of the house are practically proof against the passage through them of air, but the window and door casings and the ceilings, need equally efficacious treatment, and some form of cement or plaster is needed in each case. For several months past the writer has been conducting a series of experiments in this line, and notwithstanding the apparent simplicity of the undertaking, the most annoying obstacles were constantly met with and required great ingenuity to overcome. The substances experimented with were common mortar, Portland cement, Trinidad Asphalt, plaster of Paris, and a plaster called Adamant. The experiments were begun with galvanized iron boxes. One face of the boxes was covered with wire netting, sunk three-quarters of an inch below the top of the box, and this netting was covered with the above materials, except in the case of the asphalt, a thickness of one-fourth of an inch of this material being deemed sufficient for the experiments, and this layer was put over a half-inch of mortar. These boxes furnished practically the conditions of general plastering work. They were allowed to dry four months, when all but the Adamant were found so cracked and so shrunk as to leave wide cracks at the edges, and this line of experiment was abandoned. It may be that the use of mortar with the asphalt caused the cracks in the latter. Glass tubes were next used, and an inch of each material was placed in a tube, and this experiment also failed, as all the substances except the Adamant shrunk from the glass. and it was found impossible to stop up the spaces between the substance and the glass. In this case the substances were dried one month in a warm room. The next line of work was a success. Circular disks about two inches in diameter were cut from the materials used in the first experiment, and in each case disks free from apparant flaws were obtained. These were placed in tin tubes of larger diameter, and the annular space between the tube and the disk was cemented up with a specially prepared cement, into which a little asbestos was mixed.

These tubes were soldered onto smaller ones in order to reduce the diameter. By the use of mercury a torricellian vacuum was produced, and with a glass tube for the bowl, a scale was obtained, and the relative porosity of the materials was approximately obtained, but it is to be observed the many cracks common to large surfaces were eliminated.

The asphalt was found to be practically air-tight, and was therefore taken as the standard and its porosity called zero, and the following approximate scale was made:

	Porosity.
Asphalt	00
Adamant	5
Portland Cement	15
Plaster of Paris	60
Mortar, three coats, including white coating	75
Common Mortar, two coats	90

The one important qualification of these conclusions is, as above observed, that they do not take note of the cracks caused by drying, for selected portions were taken in each case; but the first experiment with the galvanized iron boxes and also with the glass tubes showed that the Adamant alone did not crack from shrinkage in drying. The asphalt gives off at all times a slight and, to some people, an offensive odor, and its use for inner walls is thus prohibited, but for cellar floors, where the ground is damp, and, it may be, for outer walls, its use is to be highly recommended. The Adamant gave also the hardest surface, and dried apparently perfectly in a few days. Its cost, we believe, is not much above that of good three c at common plaster work, but upon this point we are not informed. In a house built as above, the doors and window frames cemented around with Adamant, and the same material used for walls and ceiling, the problem of the distribution of warm air is a simple one of arithmetical calculation — so much warm air admitted and so much removed by a power easily determinable; but until the room is practically air-tight, the temperature at the ceiling and at the floor will show a variation of from 20 to 70 degrees, making an absolute failure of fine essays, of fine theories, of fine apparatus, and of a fine system known as the human body.

47

ADAMANT.

Will give you a solid wall.

Is the best fire-resisting plaster.

Of itself will not crack, swell or shrink.

Will not cleave when used as directed, even in
case of leakage.

Will give you a warm house.

Does not ruin woodwork by loading it with
moisture.

Admits carpenters following plasterers in a few
hours.

Is capable of every variety of finish.

ADAMANT.

Gives a first-class wall at moderate expense.

Always ready to use in any season.

Does not hold gases or disease germs.

Is the par excellence for patching.

Has received the Medal of Excellence from American Institute of New York.

Adamant is no experiment.   It has bad four years' test.

Has been applied to thousands of buildings.

Is the standard first-class wall plaster.

Adamant Companies.

NEW ENGLAND ADAMANT CO., 21 FEDERAL ST., BOSTON, MASS.

ADAMANT MFG. CO., 309 Genesee St., Syracuse. N. Y.

KEYSTONE PLASTER CO. 233 North 23d St., Philadelphia, and Builders' Exchange, Pittsburgh, Pa.

CONNECTICUT ADAMANT WALL PLASTER CO., 460 Grand Ave., New Haven, Conn.

NEW JERSEY ADAMANT MFG. CO., 61 Passaic Ave., Harrison, N. J.

OHIO PLASTER CO., 5 Euclid Ave., Cleveland, O.

ADAMANT WALL PLASTER CO., 17 and Q Sts., South Omaha, Neb.

NORTHWESTERN ADAMANT MFG. CO., 4th Ave., near 1st Street, Minneapolis, Minn.

UNITED ADAMANT PLASTER CO., 10 South Holliday Street, Baltimore, Md.

COLORADO ADAMANT MFG. CO., Denver, Col.

INDIANA ADAMANT WALL PLASTER CO., Indianapolis, Ind.

MICHIGAN ADAMANT WALL PLASTER CO., Detroit. Mich.

ST. LOUIS ADAMANT PLASTER CO., 6 to 16 North 2d Street, St. Louis, Mo.

ADAMANT PLASTER MFG. CO., Tacoma, Washington.

THE TENNESSEE ADAMANT PLASTER CO., Nashville, Tenn.

SOUTHERN ADAMANT PLASTER MFG. CO., Norfolk, Va.

SOUTHEASTERN PLASTER CO., Savannah, Ga.

OHIO ADAMANT PLASTER CO., 58 Board of Trade, Columbus. O.

TOLEDO ADAMANT WALL PLASTER WORKS, 315 Superior Street, Toledo, Ohio.

CHICAGO ADAMANT PLASTER CO., 517 Chamber of Commerce, Chicago, Ill.

CALIFORNIA ADAMANT WALL PLASTER CO., San Francisco, Cal.

ADAMANT MFG. CO., 100 Esplanade Street E., Toronto, Ont.

ADAMANT CO., LIMITED, 105 Colmore Row, Birmingham, Eng.

THE N. S. W. ADAMANT MFG. CO., 54 Margaret Street, Sydney, Australia.